Queen o the Moon

Written by Liz Miles

Illustrated by Alida Massari

Collins

The moon has lots of rock.

The queen is fed up with it.

She moans.

Boiling hot rocks shoot by her hair.

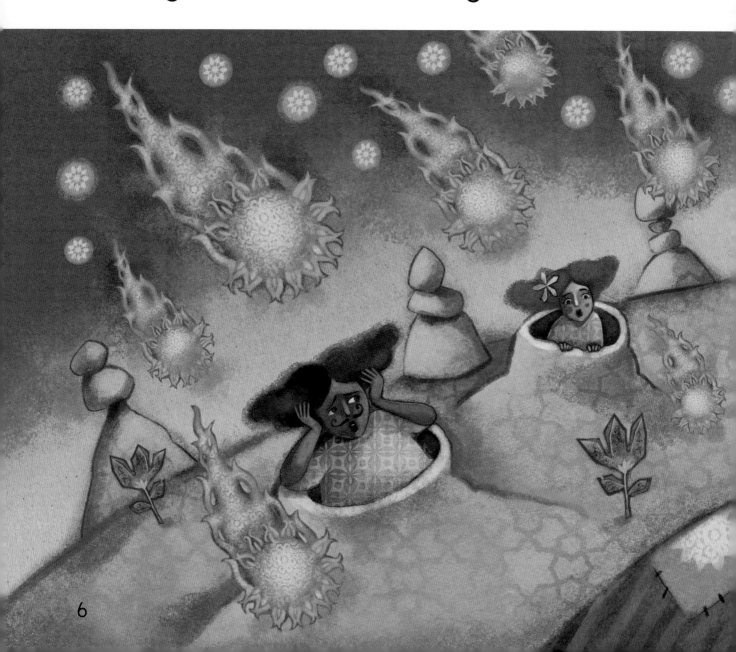

7

8

11

Now, she *never* moans.

Moon seeks queen!

❧ Review: After reading ❧

Use your assessment from hearing the children read to choose any GPCs, words or tricky words that need additional practice.

Read 1: Decoding

- Turn to page 6 and discuss what **boiling hot rocks** on the moon might be. Talk about meteorites and how they come from outer space, and can heat up as they fall through the air.
- Call out each page number below and challenge the children to find and read a word with these sounds:

page 2: /oo/ (*moon*)	page 3: /ee/ (*queen*)	page 4: /oa/ (*moans*)
page 5: /ur/ (*hurts*)	page 6: /oi/ (*boiling*)	page 10: /ar/ (*park*)

- Challenge the children to read pages 2 and 3 aloud and fluently, blending the words in their head (silently).

Read 2: Prosody

- Focus on the queen's speech bubbles from pages 8 and 9.
- Discuss what sort of the voice the queen might have. (e.g. *posh*)
- Discuss how she is feeling on each page. (page 8: *cross/determined*; page 9: *excited/curious*)
- Challenge the children to read each speech bubble expressively, to show the change of feeling.

Read 3: Comprehension

- Ask the children whether they have read any stories or rhymes about the moon. Encourage them to compare the characters and settings with the queen and her rocky home.
- Pose the question: What is the queen's main problem? Discuss why she leaves the moon and what she gives up. (*she is fed up with the rocks; she gives up her job as queen*)
- Look together at the advert for a new queen of the moon on pages 14 and 15.
  - Discuss what the job offers and ask: Would you like the job? Why? Why not?
  - Discuss why the queen left the moon and what she might miss.
  - Talk about whether the queen was right to leave the moon. Ask: Should she have stayed? Why?